TUGS AND TOWING AROUND BRITAIN

IAN COLLARD

AMBERLEY

First published 2024

Amberley Publishing
The Hill, Stroud
Gloucestershire, GL5 4EP

www.amberley-books.com

ISBN 978 1 3981 1654 2 (print)
ISBN 978 1 3981 1655 9 (ebook)

British Library Cataloguing in Publication Data.
A catalogue record for this book is available from the British Library.

Origination by Amberley Publishing.
Printed in the UK.

Introduction

Tugs are described as vessels that manoeuvre ships by pushing or towing them. They are powerful for their size and are strongly built. Some are employed as oceangoing while some serve as salvage vessels, and modern tugs are fitted with diesel engines. Many have specialised monitors fitted that allow them to assist in firefighting, particularly in harbours and for assisting vessels in emergency situations.

The main role of the tug is to assist vessels either in the river or within the dock system by moving them through pushing or towing. They are also used to tow barges or platforms that have no engines or methods of propulsion. Tugs are designed to be highly manoeuvrable and powerful as they normally work with large vessels in restricted spaces, such as docks or narrow channels. Tugs are employed to work with passenger liners, cargo vessels, oil tankers and a variety of vessels working on the river. The modern tug is equipped with azimuthing thrusters or Voith Schneider Propellers, which enable them to generate the thrust required for towing the larger vessels that are now being built for the various trades and services provided by shipowners.

The majority of vessels described in this book are harbour tugs that assist in berthing and un-berthing vessels in the various dock systems and rivers. This is a vital function and part of the work in all modern ports, which would not be able to function without the assistance offered by the modern diesel tug. It has been claimed that the *Charlotte Dundas* was the first tug as it towed two loaded barges along the Forth and Clyde Canal in 1802. Steam-powered tugs were recognised by shipowners when their vessels were delayed because of high winds, other weather or tidal-related issues. Paddle steamers were driven by low-pressure boilers which supplied steam to the side lever engines. In the second half of the century tugs were built with two engines side by side. This enabled the paddles to be controlled independently and gave the vessel a great deal of manoeuvrability. *Francis B. Ogden* was the first screw tug, followed by the deep-sea vessel the *Archimedes*. Compound engines gave greater power and better fuel consumption and enabled the tug owners to increase their profits.

The famous painting of the fighting *Temeraire* showed the *Monarch* towing the vessel to Beatson's ship breakers at Rotherhithe. *Monarch* was owned by William Watkins, which was established in 1833. The company also owned the *Anglia*, which towed *Cleopatra's Needle* on part of its voyage from Egypt to the Thames. Paddle-wheel vessels were replaced by screw-propelled tugs, and the introduction of Kort nozzles made the steam-driven tug more manoeuvrable. Around the middle of the nineteenth century the inverted steam reciprocating engine was introduced in the single-expansion form. Compound or double-expansion engines were introduced around the 1860s, with triple-expansion engines that combined high and low pressure fitted to new tugs from 1875.

In 1856, James Steel from Greenock acquired several lighters from the Clyde Shipping Company and began to trade as James Steel & Sons. This company incorporated the Glasgow and Greenock Shipping Company. They entered into the lighterage trade between Glasgow and Greenock, where the oceangoing cargo vessels lay at anchor. A subsidiary company was formed in Londonderry in 1872 under the name of Steel and McCaskill, with J. W. Currie as manager; this was changed in 1877 to Steel and Bennie. At this time the Clyde river channels were being widened and deepened to allow large vessel to navigate and dock at Glasgow. The company prospered during the twentieth century and were involved in a number of salvage operations over the course of the First and Second world wars, and in the years in-between. The *Warrior* assisted at the salvage operations of the submarine *Thetis* late in July 1939 following its sinking on trials in Liverpool Bay. Clyde Shipping Company Limited and Steel and Bennie were amalgamated under one control at the outbreak of the Second World War. Their combined fleets were later augmented by tugs from the Thames, Holland and Belgium.

Lawyer George Bell Cowl opened his tug business in 1882. He was operating with five second-hand vessels by 1887 when his company was taken over by the Alexandra Towing Company. They established their headquarters in Cowl's offices at Langton Dock, and the Alexandra Towing Company was established with a capital of £25,000 in £5 ordinary shares. Five tugs – *Flying Breeze*, *Flying Kestrel*, *Flying Tempest*, *Flying Whirlwind* and *Turbot* – were taken over from George Bell Cowl. He was invited onto the board of directors and was given 400 shares in the company. The Mack family held the most significant number of shares. The *Alexandra* was purchased in 1888. It took its name from the new company, but also from one of Liverpool's docks, and this started the fashion of naming their vessels after Liverpool docks. When they later expanded to other ports this system was adhered to with local names. The company was well managed and they were able to keep revenue up and costs down, but in one month in 1889 three tugs were involved in five accidents and a reserve fund was set up. Tug masters and engineers received a half-yearly bonus as an incentive if their vessel had not been involved in any incidents.

The master mariners William and Thomas Jolliffe founded their towing company in 1854 with their first paddle tug, the *Lioness*. It was followed in 1859 and 1860 by the *Lion* and *Emperor* and their vessels were also involved in the excursion trade to North Wales in the summer months. *Great Emperor* and *Great Western* were introduced in 1864 and were followed by the screw tug *Thomas Jolliffe* in 1879. *William Jolliffe* joined the fleet six years later and was sold to Canadian Pacific Steamships in 1914, where it was renamed *Nitinat* and used for barge towing and salvage duties. In 1924 it was purchased by the Pacific Salvage Company, becoming *Salvage Chief*, and was wrecked on Merry Island on 7 February the following year. The two-funnelled *Jane Jolliffe* entered service in 1888 and *Sarah Jolliffe* two years later. The *Andrew Jolliffe* was built at South Shields in 1894, *Vivian Jolliffe* in 1897 and *Hannah Jolliffe* in 1900. In 1908 the business of W&T Jolliffe was acquired by the Alexandra Towing Company and the *Andrew Jolliffe* was renamed *Wapping*. It was not until 1956 that a new system was inaugurated towards the naming of Alexandra tugs. As well as continuing to use the names of Liverpool docks, it was recognised that the company originated in the north end of the docks. *North Light* and *North Rock* were also completed. The motor tug *Herculaneum* was completed in 1962 with a Kort rudder. Six motor tugs followed, which were built by W. J. Yarwood at Northwich and Richard Dunston (Hessle) Limited.

Russell Rea set up his business as a coal merchant in Liverpool in 1872 and went into partnership with his brother James Rea in 1879. They established the Rea Towing Company two years later and entered into the bunkering trade with the *Cumbria*, which was later renamed *Hallgarth*. *Holmgarth* followed in 1899, *Aysgarth* in 1900 and *Fallgarth* and *Edengarth* in 1902 and 1903 respectively. *Aysgarth* was the first of their tugs to be registered in Liverpool, giving good service to the company until sold to James Dredging, Towage and Transport Company Limited in 1932. The Rea Transport Company was formed in 1902 to service the coaling activity in the Mersey, delivering fuel by barge to vessels anchored at the Sloyne. *Fellgarth* and *Edengarth* entered service and spent most of their careers towing Rea's barges on the Mersey and the

Manchester Ship Canal. In 1919, both tugs were transferred to Rea Limited and to the Rapid Coaling Syndicate in 1947, and were broken up by William Cubbin Limited at Birkenhead and Routledge Brothers, Garston, in 1960.

In 1896, the Johnston Line ordered a tug from A. W. Robertson Limited of London to assist with the docking arrangements for its deep-sea vessels in the Mersey. The original ship was named *Amore* and it worked in conjunction with vessels of other companies. It was requisitioned by the Admiralty in 1916, re-boilered and then based at Queenstown. The Johnston Line was taken over by Furness, Withy & Company Limited in 1916 and was joined by the tugs *Beemore* and *Ceemore* in 1929. *Amore* was replaced by *Deemore* the following year. *Beemore* was sold to the Alexandra Towing Company in 1958 and renamed *Mumbles*. Later that year Alexandra's purchased the *Deemore*, which became the *Margam*. *Ceemore* was also acquired in 1959 and was renamed *Murton*. The three vessels were originally employed at Liverpool but were later transferred to work at Swansea. *Murton* was broken up at Briton Ferry in 1964, *Margan* at Silloth in 1965 and *Mumbles* at the River Neath. *Rossmore*, *Foylemore* and *Kilmore* were delivered in 1958 and were painted with the name 'Furness Lines' on their hulls three years later. The three tugs were sold to Rea Limited in 1969, becoming *Rossgarth*, *Foylegarth* and *Kilgarth*.

Queen Victoria opened the Manchester Ship Canal from the decks of the Admiralty yacht *Enchantress* on 21 May 1894. The queen embarked at Latchford Locks and the yacht was towed along the canal by the tug *Florida*, which was built at Leith in 1887 and purchased by the canal company in 1891. Paddle tugs were used on the canal, and a number of narrow boats were built and fitted with steam engines for use on the Bridgewater system. The tugs *Bridgewater*, *Earl of Ellesmere*, *Brackley*, *Helen*, *Dagmar*, *Alice*, *Gower*, *St Winifred* and *Queen of the Mersey* were acquired, and Liverpool tugs were also initially licensed to operate on the canal. *Mercia*, *Charles Galloway*, *Eva*, *Hilda*, *Minnie*, *Gwennie*, *Pomona*, *Eastham*, *Partington* and *Agnes Seed* were purchased to tow and assist vessels using the canal. The paddle tugs *Barton* and *Irlam* were built at South Shields to work on the waterway, followed by four similar tugs. The *Ralph Brocklebank* was built in 1903, and from 1929 this replaced the *Charles Galloway* as the canal tender. In 1936, it was completely refitted and renamed *Daniel Adamson*, the name of the first chairman of the Manchester Ship Canal Company. From 1926 the company prefixed all names of tugs with the letters 'MSC'.

In 1916, James H. Lamey established a towing company. James and his brother and partner William had been pilots on the Manchester Ship Canal and purchased the paddler *Hero* under the name of the Hero Tug Company, managed by Marwood and Company. The Lamey family came from Appledore, Devon, where they had been owners of coastal sailing vessels, some of which carried clay to Runcorn, where it was loaded into barges and sent to the Potteries by canal. Lamey also owned the schooner *Irish Minstrel* and the auxiliary schooner *Puseyjones No 2*. The *Hero* was built in Bristol in 1896 and was purchased by the London Grain Elevator Company in 1898. It was sold to John Ryan of Birkenhead in 1906 and was acquired by Lamey in 1916. The name of the business was changed to J. H. Lamey & Company in 1920.

The West Coast Towing & Salvage Company was set up in 1917, taking over the Star Tug Company, the Steam Tug Expert Company and the Sunshine Tug Company. However, they only survived until 1924. In 1919, when the Cunard Steamship Company moved some of its operations to Southampton, the Alexandra Towing Company also established a base at that port. It was at this time that investment in R&JH Rea was encouraged to enable the company to acquire new ships and infrastructure following losses during the First World War. Rea Limited was then formed, with Alfred Holt & Company as a major investor. A takeover offer was received from William Cory & Son Limited, who then purchased the Rea Transport Company, which included interests outside the River Mersey. William Cory acquired 50 per cent of Rea Limited, and the Rea Transport Company Limited was restyled as R&JH Rea Limited in 1919.

Oil-driven vessels were introduced during the 1950s, replacing steam-powered tugs in the ports around the United Kingdom. The new engines provided fuel economy, had a greater power-to-weight ratio, and reduced machinery space and operational availability. A number of steam tugs were converted to oil-burning vessels during this period. The new tugs were installed with larger engine plants, which provided greater horsepower and heavier bollard pull rates. This additional space was provided because the diesel tugs did not require as many crew to operate them as the older steam-driven vessels.

The change in tug propulsion coincided with the revolution in shipping, which saw the introduction of new large container vessels and oil tankers. By the early 1960s the transportation of goods in containers was gaining momentum and the large shipping companies were discussing plans to build purpose-built container ships. The twenty-year period between 1960 and 1980 was clearly a significant time of change for all those involved in the maritime and shipping industries. The advantages of the carriage of goods in containers was realised and investment was made in new ships, ports and other transport facilities which required the disposal of many conventional vessels and ports.

The Thames tug *Hibernia*, owned by William Watkins Limited, was delivered to T. W. Ward at Grays, Essex, and was broken up in 1961. It was managed by Ship Towage (London) Limited and was built at Delfshaven by Maats de Maas in 1884 as an oceangoing tug. In 1902 it towed a paddle steamer to the Persian Gulf in forty-two days, beating a Dutch tug with a similar tow by eleven days. It was renamed *Hibernia III* in 1905 and served at the Dardanelles during the First World War. It became *Hibernia* in 1919 and had 15 feet taken from its bow in 1924 to make it suitable for work on the River Thames.

The Rea tug *Kingsgarth* sank outside the lock at Avonmouth on 3 February 1964 following a collision with the cargo vessel *Port Launceston*. Three members of the crew were lost and salvage work was undertaken by the Liverpool and Glasgow Salvage Association. It was floated and arrived at Newport on 12 July 1965, where it was broken up. On 15 February 1968, the Bristol steam tug *Bristolian* sailed from Avonmouth to John Cashmore at Newport to be broken up after fifty-seven years of service in the Bristol Channel. Its final duty was assisting the *New York City* on 17 December 1967. *Bristolian* was built for C. J. King and Sons Limited with a large forward cabin that had teak and mahogany furniture. It had a certificate for 199 passengers and often acted as a tender for the passenger liners *Royal George* and *Royal Edward*, which operated between Avonmouth and Canada prior to the First World War. In its many years of service *Bristolian* only had four skippers: *Captains Bill Allen*, *George Havens*, *Bill Bevan* and *Frank Burton*. Its first engineer, Frank Jenkins, was with it for ten years. He was succeeded by Fred Bussey, who held the post for over thirty years. On 8 November 1912, it assisted the steamer *Winchester* when it ran aground off Sharpness. *Bristolian* also assisted in the salvage of the sailing ship *John Ena*, which was dragging its anchor in a strong gale and was saved from going ashore. It helped to re-float the *Sacramento Valley* from a sandbank in the Severn in 1936. The *Henri Jaspar* and *Tafelberg* were mined off Barry in 1940 and were in danger of sinking. They were both beached by *Bristolian* and later repaired and returned to service.

Two of the main tug-owning companies on the Thames merged their activities in 1968. William Watkins Limited Ship Towage (London) Limited and W. H. J. Alexander Limited then had a total fleet of thirty-four tugs; the new company was known as London Tugs Limited. Ship Towage (London) Limited acted as managers for William Watkins Limited, incorporating the Elliott Steam Tug Company and Gamecock Tugs Limited.

Shipping companies and other maritime agencies that had been operating successfully for over 100 years needed to amalgamate in order to be able to finance the larger vessels and dock facilities required to operate services. The availability of cheaper air travel was a major factor in the decline of passenger travel by sea, with some shipping operators diversifying by providing cruises with their ships and others moving out of the market completely. Consequently, the demands on the industry were drastically reduced as many of the smaller tugs were sold and orders were placed for vessels capable of assisting the larger container ships and oil tankers which were using the new facilities being provided at ports around the United Kingdom.

In 1972, Cory acquired the Grangemouth & Forth Towing Company, the Clyde Shipping Company and W. H. Reynolds of Plymouth the following year. The *Waterloo*, *Wellington*, *Wallasey*, *Sun London*, *Sun Essex* and *Sun Kent* were delivered. A share of ITM Offshore Limited was acquired in 1973, and the Alexandra Marine Transportation Company was formed to oversee the operation of Alexandra's offshore services. The grain business Bulk Cargo Handling Services in Liverpool was closed in 1985 and the staff made redundant. *Indomitable* and *Formidable* were delivered in 1978–79, *Canada* in 1980, *Collingwood* in 1981, and *Sun Thames* and *Ganges* in 1982. The States of Jersey ordered a multipurpose tug in 1971 from J. Bolson and Son at Poole, Dorset. It was named *Duke of Normandy* and the main engine was a Caterpillar diesel of 565 bhp, controlled from the wheelhouse, which had a bollard pull of over 7 tons. It was designed to work mainly in the port of St Helier and to act as a tender, maintaining buoys and light beacons or carrying thirty-six passengers. It was also designed to act as a firefighting vessel to spray detergent on oil slicks and salvage pumping when required.

The *Wapping* sailed from Swansea in 1973 to undertake towage duties in Gibraltar. The following year the Alexandra Towing Company acquired London Tugs, which became Alexandra Towing Company (London) Limited. The Medway Dry Dock and Engineering Company at Sheerness and the Britannia Marine Towing Equipment also came under its control at this time. Bulk Cargo Handling Services and the thirty-one barges of the Liverpool Lighterage Company were acquired by Alexandra in 1971. The following year they also bought the barge fleet owned by Rea Towing Limited and one of their barge tugs. Bulk Cargo Handling Services purchased the seven elevators of the United Grain Elevators in 1973 and ordered three new grain barges. *Alfred* and *Crosby* were delivered in 1971, and *Albert* and *Victoria* the following year. R&JH Rea took over Steel & Rennie of Glasgow and the Newport Screw Towing Company in 1970. R&JH Rea Limited was restyled as Cory Ship Towage Limited in 1970 and the fleet, except that of the Rea Towing Company on the Mersey, was repainted in the new livery.

In 1982, the North British Maritime Group reported an upturn in their towage subsidiaries United Towing Limited and Humber Tugs Limited. The *Salvageman* and *Yorkshireman* remained in service with the Ministry of Defence and *Irishman* and *Euroman* operated in commercial deep-sea operations, plus a number of vessels from Humber Tugs Limited. The *Masterman* was upgraded for use in short sea and coastal work, joining the *Lady Debbie*, *Lady Moira* and *Lady Vera*. Humphrey and Grey Lighterage Limited ceased trading as one of the established major lighterage operations on the Thames. The company operated five tugs, mainly in the grain trade, from Tilbury Grain Terminal. C. J. King and Sons Limited and Cory Ship Towage Limited merged in 1983 to form Cory-King Towage with seven vessels in the combined fleet. The tug owners A. E. White and Son Limited ceased trading in 1983. They were involved in contracting work for civil engineering projects, dredging and general towage. The Clyde Shipping Company (Tugs) Limited purchased the interests of Lawson-Batey, who controlled Tyne Tugs Limited of Newcastle, France Fenwick, Tyne and Wear Limited of Sunderland and the Blyth Tug Company. Clyde Shipping were associated with Shetland Towage Limited and Forth Tugs Limited of Grangemouth.

At the beginning of 1983 Holyhead Towing Limited acquired two tugs for their fleet. *Karet* was built by Richards Shipbuilders at Lowestoft in 1967 and was purchased from Shell International Marine. It had been used for tanker handling and was also designed for use as a firefighting vessel. It was constructed with a General Electric main engine of 2,250 bhp and was renamed *Afon Goch*. The Ministry of Defence Girl class vessel *Alice* was also purchased as a 37-ton harbour tug. Changes were made to its steering gear, accommodation and navigational equipment, being renamed *Afon Cefni*. The two new vessels then joined *Afon Las*, *North Stack* and *Carmel Head*.

Sun Anglia was delivered in 1984 and *Bramley Moore* the following year. *Redoubtable* was purchased in 1982, and *Implacable* and *Invincible* the following year. In 1983, Rea's took over C. J. King of Avonmouth. Alexandra Marine Transportation Limited was an offshore division of the Alexandra Towing Company and was structured as a marketing company for anchor-handling tugs, supply vessels and flat-top pontoon barges.

Tugs need to be highly manoeuvrable, so a number of different propulsion systems have been developed to increase manoeuvrability and safety. The early tugs that were driven by paddle-wheels were replaced by propeller-driven tugs; Kort nozzles were added to increase thrust, and the nozzle-rudder omitted the need for a conventional rudder. Prior to the Second World War the cycloidal propeller was developed and fitted to tugs to increase their manoeuvrability, while after the war the Voith Water Tractor was developed. The azimuth thruster was developed in the 1950s. These propulsion systems are used on tugs designed for ship docking.

The Kort nozzle is a cylindrical structure fitted around a special propeller; there is a minimum clearance between the propeller blades and the inner walls of the Kort nozzle. The thrust-to-power ratio is enhanced as the water approaches the propeller in a linear configuration and exits the nozzle the same way. The nozzle is named after its inventor but there are many different brands.

The *Carousel Tug* was the winner of the Maritime Innovation Award at the Dutch Maritime Awards Gala in 2006. The tug includes a pair of interlocking rings to its body. The inner ring is attached to the vessel, with the outer ring attached to the towed ship by winch or towing hook. The towing point then rotates freely, making it difficult to capsize the tug.

The Voith Schneider propeller, or cycloidal drive, is also a marine propulsion system. It enables the vessel to be very manoeuvrable, allowing it to change direction almost instantaneously. It is fitted to a circular plate that protrudes from the bottom of the ship. Each blade rotates around a vertical axis. The internal gear changes the angle of the blades in sync with the rotation of the plate. Each blade provides thrust in any direction, very similar to the collective pitch control and cyclic in a helicopter.

Diesel engines in tugs typically produce 500 to 2,500 kW (680 to 3,400 hp), but larger vessels, which are used in deep waters, can have power ratings up to 20,000 kW. Tugs are typically rated by their engine's power output and overall bollard pull. The bollard pull of a tug refers to the static force exerted by the tug pulling on a fixed object during a test in deep water. The force on the towline is measured and used as a standard for the towing capability of the tug.

Tugs play an important part in the operations of the ports around the United Kingdom. Their duties involve:

- Berthing and unberthing of vessels
- Pilotage services
- Subsea inspection
- Repair and maintenance
- Firefighting
- Escorting vessels
- Crew and cargo transportation
- Coastal and deep-sea towage

Members of the British Tug Association (BTA) operate more than 150 tugs in over forty ports around the British Isles, employing more than 1,000 seafarers. One of the main aims of the organisation is to improve safety in the industry by encouraging best practice and ensuring that all seafarers receive the appropriate training and certification. The association has a close relationship with the United Kingdom Chamber of Shipping; tug-operating members are full members and are represented in the full range of committees, panels, seminars and networking events. This strengthens the voice of the association with the government when matters involve regulatory development with European partners and the International Maritime Organisation (IMO). The BTA was founded in 1934 and their members' tug fleets cover ports handling 90 per cent of Britain's seaborne trade, including all of the leading ports. Membership allows tug operators to be consulted over new legislation, training, health and safety issues, port and pilot liaison, vessel design and education.

In complying with the Port Marine Safety Code (PMSC), the Statutory Harbours Authority (SHA) identified towage as a major factor in reducing risk in certain shipping operations. Consequently, the SHA requires an adequate number of licensed tugs to be on duty within a harbour area. The tugs must be 'fit for purpose' with qualified crews and competent pilots trained to agreed standards. Towage companies are required to have their tugs licensed by the SHA, and these licences are renewed annually. All crew members must meet certain requirements, and the tugs must be adequately manned.

The Orkney Towage Company took delivery of *Erlend* and *Einar* in 1991, which replaced two conventional twin-screw tugs, *Kessock* and *Kinloch*. *Erlend* and *Einar* are very advanced stern-drive tugs designed by Bureau Wijsmuller and constructed by McTay Marine Limited at Bromborough. They are powered by two Ruston main engines of 3,860 bhp and twin azimuthing propulsion units, and have a bollard pull of over 50 tons. The new tugs work alongside the *Kintore*, which is a twin-screw vessel of 3,960 bhp, built for J. P. Knight in 1977. When it joined the Orkney fleet it was fitted as a fire tug. J. P. Knight have a financial interest with Orkney Towage. *Kessock* and *Kinloch* were owned by Knight and were soon repositioned in British Guiana. They were modified for use as pusher tugs to work with barges carrying 6,000 tons of bauxite 64 miles from the mines to New Amsterdam. Knight's *Kindeace* and *Keverne* are also involved in the contract to tow the barges 94 miles from New Amsterdam to Georgetown. J. P. Knight Limited acquired Goliath BV, the Dutch heavy-lift company, in 1991, forming the subsidiary Goliath-Knights BV. Caledonian Towage & Marine Services Limited was another subsidiary of Knights and this became J. P. Knight (Caledonian) Limited in 1991, with Kindeace being transferred to the contract in Guiana. The company experienced significant growth and the Japanese *Kuroshio* was purchased, becoming *Kestrel*. In June 1991, J. P. Knight (Rochester) Limited handed over their Medway operations to Howard Smith (UK) Limited to operate it as Medway Towage Limited, a subsidiary of Humber Tugs Limited. This sale did not affect the other J. P. Knight operations in the offshore market, Scotland and overseas, and involved five tugs and thirty-five people.

Humber Tugs placed a £15 million order for five tugs with McTay Marine in 1989. The company was a subsidiary of Howard Smith (UK) Limited. All five vessels were of the same design, with the first vessel, *Lady Anya*, being delivered on 14 December 1990. It is a tractor tug of the Voith Schneider type, being 30 metres in length with a breadth of 11 metres. The master is able to control the propulsion system and they are fitted with soundproofing in the crew accommodation areas. Fire monitors are mounted above the wheelhouse and supplied to pumps driven from the main engines. These are capable of delivering 20,000 litres of water per minute. Ruston 6RK270M diesel main engines are installed, with each producing in excess of 2,300 bhp at 1,000 rpm. The engines are coupled to the Voith Schneider propulsion units via turbo couplings, giving a bollard pull of 56 tons and a speed of 12.8 knots.

The Alexandra Towing Company, with its operations in Liverpool, London, Felixstowe, Southampton, Swansea and Gibraltar, was taken over by the Australian Howard Smith Group in 1992 when a bid of £52 million was accepted. However, it was not until spring 1994 that the Alexandra livery finally disappeared from the entire fifty-three-vessel fleet and the rectangular Howard Smith logo appeared on the superstructure. The following year Howard Smith Towage & Salvage announced an order for six new tugs: two for Felixstowe, one to Southampton and one for use by the Medway Towage operation at Sheerness. Options would be taken on two further vessels.

In May 2001, the towage operations of Howard Smith were acquired by Adstream Marine Limited, becoming Adstream Towage Limited and based at Huskisson Dock, Liverpool, with its head office in Hull. In 2007, Smit Internationale NV entered the United Kingdom tug market when it acquired Adstream as part of its Liverpool operations. On 14 January 2000, Cory Towage Limited was sold to Wijsmuller for £81.8 million, becoming part of the Svitzer Marine Group in 2006.

The first fully electric tug entered service in Istanbul in September 2020. The *Gisas Power* is owned by Gisas Shipbuilding and operates in the Port of Tuzla Aydinli Bay. Its power is generated by two 1,450 kW battery packs, which are located separately forward and aft, and are air cooled. Experiments are currently taking place with more powerful electric vessels and also with tugs that are propelled by hydrogen.

Thames

Atlantic Cock, 1932, 182grt, 26.3×7.6, 1000 (1) SR, IMO 1162687.
b. A. Hall Limited, Aberdeen Yard, No. 639.
William Watkins (Ship Towage) Limited.
It was broken up in 1970.

Ocean Cock, 1932, 182grt, 26.3×7.6, 1000 (1) SR, IMO 5260227.
b. A Hall Limited, Aberdeen Yard, No. 640.
William Watkins (Ship Towage) Limited.
It was broken up in 1969.

Napia, 1943, 261grt, 1200 (1) SR, IMO 5247134.
b. Goole Shipbuilding Company, Goole Yard, No. 398.
William Watkins (Ship Towage) Limited.
Built as *Empire Jester*, becoming *Napia* in 1946 and *Tolmiros* in 1971. Broken up in 1986.

TID-172, 1946, 54grt, 19.8×5.1.
b. Henry Scarr Limited, Hessle.
TID-172 was allocated to the Nore Command under the naval officer in charge at Lowestoft and named *Martello*. It reverted to the name *TID-172* on 11 July 1946 with a civilian crew. On 20 October 1946, it was transferred to the civil engineer-in-chief at Chatham and became *W92*. On 1 October 1959, it was transferred to the Port Auxiliary Service at Chatham, becoming *TID-172*. Sold for demolition on 9 July 1973 to T. W. Ward Limited but was later saved for preservation. In December 2020, it was located at Hoo, Rochester.

Crested Cock, 1935, 177grt, 29.3×7.6, 1000 (1) SR, IMO 5082039.
b. A. Hall Limited, Aberdeen Yard, No. 654.
William Watkins (Ship Towage) Limited.
It was broken up in 1970.

Ionia, 1960, 187grt, 1350 (B), IMO 5162449.
b. Henry Scarr Limited, Hessle, Yard No. 771.
William Watkins Ship Towage (London) Limited.
It was sold and renamed *St Mawes* in 1988 and *Ionia* in 2005.

Crowstone (1932/48grt), *Danube VII* (1946/237grt), *Danube VIII* (1946/237grt) and *Danube* VI (1935/241grt).

Racia, 1930, 163grt, IMO 5288786.
b. Cochrane & Sons Limited, Selby, Yard No. 1091.
William Watkins Ship Towage (London) Limited.
Built as *Dilwara*, becoming *Dendera* in 1935 and *Racia* in 1938. Broken up in Belgium in 1967.

Sun Thames (1982/330grt), *Sun Essex* (1977/272grt) and *Sun III* (1966/150grt).

Sun VIII, 1919, 196grt, 32.3×7.8, IMO 5344310.
b. Cochrane & Sons Limited, Selby, Yard No. 731.
W. H. J. Alexander Limited (Sun Tugs).
It was converted to oil burning in 1960 and was broken up at Antwerp in 1969.

S. A. Everard, 1939, 124grt, 800 (B) M.
Owned by F. T. Everard who also operated the smaller launch tugs *Jester* and *Joker* along with a fleet of coastal vessels.

Frisian Down (1964/80grt) was owned by Humphrey & Grey (Lighterage) Limited and is shown on towing duties in the Thames in July 1971.

Hibernia, 1963, 293grt, 2250 (I), IMO 5405085.
b. Cochrane & Sons Limited, Selby, Yard No. 1483.
William Watkins Ship Towage (London) Limited.
It was renamed *Atrotos* in 1987, *Karapiperis* in 1991 and *Alfios* in 2007.

Burma (1967/166grt) and *Sun III* (1966/150grt).

General VI was owned by Thames & General Lighterage Limited and was participating in the Thames rowing race at Blackfriars on 26 June 1976.

Blackboys (1958/22grt) was also owned by Thames & General Limited and is shown on towing duties on the Thames.

Dhulia, 1959, 272grt, 34.6×8, 8, IMO 5089465.
b. Henry Scarr Limited, Hessle, Yard No. 762.
William Watkins Limited.
It became *Dhulia S* in 1983 and was broken up in 1985.

Plasma, 1965, 122grt, 1600 (B), IMO 6504371.
b. Richard Dunston, Hessle, Yard No. s807.
Port of London Authority.
It was renamed *Burma* in 1991, *Langland* in 1994, *Lady Joan* in 1998 and *Joan* in 2001.

Sun Anglia, 1985, 339grt, 33.8×10, IMO 8415146.
b. McTay Marine, Bromborough, Yard No. 59.
London Tugs Limited (Alexandra Towing Company Limited).
It was renamed *Adsteam Anglia* in 2006 and *Svitzer Anglia* in 2008.

Sun XXVI, 1965, 230grt, 35.4×8.7, IMO 6500387.
b. Charles D. Holmes, Beverley, Yard No. 992.
W. H. J. Alexander Limited (Sun Tugs).
It became *Saga Moon* in 1997.

Sun Kent, 1977, 272grt, 32.9×9.6, 2070 (35), IMO 7531292.
b. Richard Dunston Limited, Hessle, Yard No. 911.
W. H. J. Alexander Limited (Sun Tugs).
It was renamed *Montrado* in 2001, *Stadt Tug* in 2006 and *Sharon 1* in 2013.

Sun XXIV, 1962, 113grt, 26.9×7.0, 1000 (I), IMO 5344437.
b. James Pollock & Sons Limited, Faversham, Yard No. 2127.
W. H. J. Alexander Limited (Sun Tugs).
It was sold and renamed *Kingston* in 1992.

Port of London Authority tug *Plagal* (1951/159grt).

Smit London, 1975, 2,273grt, 74.8×15.8, IMO 7402453.
b. De Merwede S&M, Hardinxveld, Yard No. 611.
Smitwijs London CV.
It became *Smitwijs London* in 1998, *London* in 2007 and *Global Change* in 2013. Broken up at Alang in 2013.

Avenger, 1975, 299grt, 39.9×10.1, IMO 7391719.
b. Tito-Sava, Macvanska, Yard No. 922.
Alexandra Towing Company.
Built as *Sea Diamond*, becoming *Chambon Bora* in 1980, *Indefatigable* in 1984, *Avenger* in 1986, *Valiant Nader* in 1994 and *WB-Force 1* in 2004.

Rochester

Keverne, 1960, 260grt, 37.5×8.7, 1650 (B) M, IMO 5186031.
b. Thornycroft, Woolston, Yard No. 4189.
J. P. Knight Limited, Rochester.
It was renamed *Vaya Condios* in 1987, *March* in 2004, *Cosani* in 2009 and *Obs 1* in 2022. Broken up in 2022 at Aliaga.

Kennet, 1965, 278grt, 34.6×9.2, 2100 (B), IMO 6510564.
b. Richard Dunstan, Lowestoft, Yard No. 480.
J. P. Knight Limited, Rochester.
It became *Linnet* in 1996 and was broken up in Holland in 2011.

Dover

Deft, 1984, 287grt, 29.9×9.0, 2800 (29), IMO 8316027.
b. McTay Marine, Bromborough, Yard No. 52.
Dover Harbour Board.
It became *Shorne* in 2000 and *HT Scimitar* in 2006.

Dextrous, 1984, 287grt, 29.9×9.0, 2800 (29), IMO 8316039.
b. McTay Marine, Bromborough, Yard No. 53.
Dover Harbour Board.
It was renamed *Cobham* in 2000 and *HT Cutlass* in 2007.

Newhaven

Foxbay (1963/151grt) and *Kingston* (1962/120grt) at Newhaven on 2 May 1997.

Meeching, 1960, 152grt, 29.3×7.7, 1320 (14), IMO 5230686.
b. P. K. Harris, Appledore, Yard No. 127.
British Rail.
It was renamed *Nore Crest* in 2013.

Portsmouth

Portsmouth dockyard tugs.

TID-32 at Pounds Scrapyard on 9 August 1988. 'TID' was a standardised British design for a tug drawn up and built during the Second World War. 182 (*TID-1* to *TID-12* and *TID-14* to *TID-183*) were built for the Ministry of War Transport.

Powerful, 1985, 384grt, 38.7×9.4, 2640, IMO 8401482.
b. Richard Dunston Limited, Hessle, Yard No. 946.
Ministry of Defence.
It was renamed *SD Powerful* in 2008.

Ministry of Defence (Navy Department) tugs at Portsmouth on 29 August 1976.

Southampton

Wyeforce (1993/57grt) and *Wyepush* (1989/50grt) owned by Itchen Marine (Towage) Limited, Southampton.

Redbridge, 1995, 399grt, 33.0×11.7, 4104, IMO 9116888.
b. Yorkshire Dry Dock Company, Hull, Yard No. 339.
Red Funnel Towage.

Clausentum, 1926, 268grt, 1200 (1) SR(2), IMO 5075787.
b. Thornycroft Limited, Woolston, Yard No. 1049.
Southampton, Isle of Wight & South of England Steam Packet Company Limited.

Canute, 1923, 271grt, 1200 (1) SR (2), IMO 5060419.
b. Thornycroft Limited, Woolston, Yard No. 1014.
Southampton, Isle of Wight & South of England Steam Packet Company Limited.

Calshot, 1930, 679grt, 45.09×10.1, 1500 (1), IMO 5058155.
b. Thornycroft Limited, Woolston, Yard No. 1014.
Southampton, Isle of Wight & South of England Steam Packet Company Limited.
It was re-engined in 1964, becoming *Galway Bay* and *Calshot* again in 1990. Dismantled and scrapped in 2022.

Calshot, 1964, 494grt, 42.4×10.7, 1848, IMO 6402717.
b. Thornycroft Limited, Woolston, Yard No. 1014.
Southampton, Isle of Wight & South of England Steam Packet Company Limited.
It was renamed *Tara II* in 1989 and *Boluda Abrego* in 1992. Broken up in 2012.

Sir Bevois, 1985, 250grt, 29.3×9.1, 2760, IMO 8414166.
b. McTay Marine Limited, Bromborough, Yard No. 58.
Red Funnel Towage.

Hamtun, 1985, 250grt, 29.3×9.1, 4200, IMO 8414154.
b. McTay Marine Limited, Bromborough, Yard No. 57.
Red Funnel Towage.

Lyndhurst, 1996, 379grt, 30.0×11.6, 4200, IMO 9129495.
b. McTay Marine Limited, Bromborough, Yard No. 117.
Red Funnel Towage.
It was renamed *Adsteam Lyndhurst* in 2006 and *Svitzer Lyndhurst* in 2009.

Lomax, 2012, 420grt, 28.2×12.6, IMO 9657832.
b. Sanmar, Istanbul, Yard No. 4.
Boreas Shipping Limited.
Built as *Sanmar Terminal VII*, becoming *Lomax* in 2013.

Sheen, 1925, 48grt, 19.5×4.5.
b. John Chambers & Company, Lowestoft.
It was converted from steam to diesel in 1962 and was owned by Tough Brothers and helped to evacuate troops from the beaches at Dunkirk during the Second World War.

Brambles (1942/242grt) and *Sir Bevois* (1953/318grt) working together at Southampton.

Poole

Kingston Lacy, 1960, 75grt, 21.0×6.0, 495, IMO 5187877.
b. Babcock Marine, Appledore.
Jenkins Marine Limited, Poole.
Built as *Kingston Buci*, becoming *Kingston Lacy* in 1984.

Wendy Ann (1934/72grt) was built by A. Hall & Company Limited at Aberdeen and was owned by Poole Harbour Commissioners Limited.

St Helier

Elektra was completed in 1953/4 as the *Groton* and *ST1982* by the Olson Group at Lake Beresford and Rawis Brothers Contracting at Jacksonville.

Duke of Normandy, 1973, 91grt, 23.6×7.0, 565 (8.5).
States of Jersey (Harbour Department).

Devonport

Christine, 1966, 77grt, 22.0×6.50, 659, Girl class.
b. Isaac Pimblott & Sons Limited, Northwich, Yard No. 817.
Royal Maritime Auxiliary Service.
It was sold in 1990 (name unchanged).

Doris (1969/77grt) at Devonport Docks on 23 August 1985.
b. Richard Dunston Limited, Thorne, Yard No. T1248.
Ministry of Defence.
It was renamed *Doris K* and *Doris* in 1990.

Plymouth

Trevol (1921/137grt), *Anthony* (1902/115grt), *Tactful* (1909/112gt) and *Carbelle* (1929/110grt).

Alsatian, 1961, 152grt, 28.7×7.4, 1320 M(2), IMO 5012577.
b. Henry Scarr Limited, Hessle, Yard No. 779.
Ministry of Defence.
It became *Clutha* in 1995 and *Cumbrae* two years later.

Valiant B, 2006, 49grt, 19×6.0, 1440, IMO 9351282.
b. Damen Shipyard Limited, Gorinchem.
Sold and renamed *St David* in 2007.

Falmouth

ST Budoc, 1958, 208grt, 31.6×8.5, IMO 5118979.
b. W. J. Yarwood & Sons Limited, Northwich, Yard No. 913.
Built as *Folylemore*, becoming *Foylegarth* in 1969, *St Budoc* in 1983, *Frazer Lloyd* in 2003 and *St Budoc* in 2009. Broken up in 2010.

St Mawes, 1951, 346grt, 800 (I) SR, IMO 5306435.
b. Henry Robb & Company Limited, Leith, Yard No. 417.
Falmouth Towage Company Limited.
It was broken up in 1984.

Gribbin Head, 1955, 132grt, 26.3×7.3, IMO 5161304.
b. Scott's of Bowland, Yard No. 408.
It was built as Ingleby Cross, becoming *Gribbin Head* in 1968, *Tuskar Rock* in 1992 and *Trivia II* in 2011. It is shown here at Fowey in July 1981.

Avonmouth

Lowgarth, 1965, 152grt, 920 (B), IMO 6504333.
b. Richard Dunston Limited, Lowestoft, Yard No. 481.
Cory King Towage Limited.
It became Charles Plane in 2006.

Sea Endeavour, 1980, 213grt, 30.5×9.0, 3150 (42), IMO 7915163.
b. Richard Dunston Limited, Lowestoft, Yard No. 549.
Cory King Towage Limited.
It became *Sea Endeavour I* in 2006 and was broken up in 2017.

Sea Merrimac, 1964, 143grt, IMO 6415104.
b. Charles Hill & Sons Limited, Bristol, Yard No. 447.
C. J. King & Sons Limited.
It became *Pullman* in 1983 and *Cantug* in 1993.

Pengarth, 1962, 160grt, 28.7×8.0, 1080 (B), IMO 5273949.
b. Charles Hill & Sons Limited, Bristol, Yard No. 438.
Cory King Towage Limited.
It was renamed *Vigilant* in 1997.

Bristol

Finch, 1956, 58grt, 19.4×5.6, 310. It was owned by Mayhew Tugs Limited, Bristol, and is shown at Bristol on 8 January 1986.

Durdham (1935/49grt) was originally *John King*, becoming *Peter Leigh* in 1970, *Pride* in 1978, *Durdham* in 1986 and *John King* again in 1988.

John King (1935/49grt) and *Mayflower* (1861/32grt) at Bristol on 16 May 2001.

Ernest Brown, 1944, 54grt, 22.3×5.2, 220.
T. R. Brown & Sons Limited.
Fitted with a diesel engine in 1964 (Rushton & Hornsby 6VCBXM, 165 KW, 225 hp).
Sank in Old Lock at Avonmouth in 1981 and later raised.

Cardiff

Bargarth, 1966, 161grt, 850 (B), IMO 6611083.
b. Richard Dunston Limited, Lowestoft, Yard No. 486.
Cory Ship Towage Limited.
It was renamed *Tennaherdhya* in 2010 and *TTMS Viking* in 2017.

Hallgarth, 1979, 204grt, 28.4×8.9, 2190 (23.5), IMO 7800057.
b. Scott's of Bowling, Yard No. 457.
Cory Ship Towage Limited.
It was renamed *St Piran* in 2008.

Barry

Goliath, 1956, 169grt, 28.7×7.6, 1290, IMO 8227214.
b. P. K. Harris Limited, Appledore, Yard No. 104.
Spithead Trading Company.
It was built as MSC *Scimitar*, becoming *Goliath* in 1987.

Uskgarth, 1966, 161grt, 850 (B), IMO 6602824.
b. Richard's Drydock & Engineering Limited, Great Yarmouth, Yard No. 484.
Cory Ship Towage Limited.
It was renamed *Tamar* in 1996.

Swansea

North Wall, 1959, 219grt, 1000 (1) SR, IMO 5257189.
b. Scott's of Bowling, Yard No. 421.
Alexandra Towing Company.
It was renamed *Maestoso* in 1973 and broken up in 1988.

Dritto, 1954, 200grt, 950 (1) SR, IMO 5386772.
b. Cochrane & Sons Limited, Selby, Yard No. 1389.
Alexandra Towing Company.
Built as *Waterloo*, becoming *Dritto* in 1972, and broken up in 1989.

Thomas, 1937, 91grt, 26.5×6.4.
b. Cochrane & Sons Limited, Selby.
Built as *Oner II*, *C10* in 1939, *Destiny* in 1958, *Affluence* in 1964, *Influence* in 1984 and *Thomas* in 1991.

Mersey

Storm Cock, 1936, 169grt, 1000 (1), IMO 5341746.
b. Cammell Laird & Company Limited, Yard No. 1017.
Liverpool Screw Towing & Lighterage Company.
It was broken up at Dalmuir in 1968.

Wachtel was built in Hamburg in 1940 and it is reported that it assisted the battleship *Bismark* from the quay when it sailed on its final voyage. During the Second World War it operated at Wilhelmshaven and became *Sheilia* at the end of hostilities. For several years it was based at Milford Haven and was later moved to Liverpool. In 1998, it sank in Sandon Dock, but was not raised until August the following year. It is seen here on the quayside at Sandon Dock, where it remained for some time until it was transferred to Holyhead, North Wales, by a preservation trust.

The Alexandra Towing Company tug *Brockenhurst* (1964/173grt) being lifted out of the water by the Mersey Dock & Harbour Board's floating crane, *Samson* (1960/774grt), following its sinking in 1984.

Hollygarth (1969/334grt) and *Cedergarth* (1962/213grtgrt) assisting *Essi Silje* off Alfred Dock, Birkenhead.

The preserved tug *Challenge* in Albert Dock. It was built in 1931 for the Elliot Steam Tug Company, London, by Alexander Hall at Aberdeen. It was converted to oil burning in 1964 and withdrawn from service with London Tugs Limited in 1971. *Challenge* holds the distinction of being the last steam tug to work on the Thames. It was bought for preservation in 1973.

Game Cock V, 1953, 218grt.
It became *Wellington* in 1970, *Vernicos Costas* in 1971, *Evangela M* in 1989 and *Kostas ii* in 1990. It was broken up in Greece in 1992.

Poderoso (1951/237grt) and *Strepitoso* (1951/237grt). *Poderoso* was built as *Formby* by Cochrane at Selby and was sold and renamed in 1969. It was deleted from the register on 16 July 2001. *Strepitoso* was also built at Selby as *Canada* and was renamed in 1969 and broken up in 1988.

Alfred Lamey, 1908, 172grt, 750, IMO 1125672.
b. Ferguson Brothers, Port Glasgow, Yard No. 176.
Built as *Darent*, it became *Alfred Lamey* in 1949 and was broken up at Barrow in 1957.

North Cock (1903/149grt) was built by Laird Brothers Limited at Birkenhead (Yard No. 656) for the Liverpool Screw Towing & Lighterage Company and was demolished at Liverpool in 1964.

Canada (1980/258grt) and *Brave Merchant* (1999/22046grt) off Langton Lock in the River Mersey.

B W Lake (2004/158557grt) is assisted by tugs in the River Mersey.

Alexandra Towing Company advertisements.

North Loch (1959/200grt) and *North Isle* (1959/200grt).

Yewgarth (1985/452grt), *Oakgarth* (1984/452grt) and *Ashagrth* (1992/307grt).

Ashgarth (1992/307grt), *Oakgarth* (1984/452grt) and *Svitzer Stanlow* (2006/656grt).

Yewgarth, 1985, 452grt, 36.9×10.5, IMO 54991.
b. McTay Marine Limited, Bromborough, Yard No. 56.
Wijsmuller Marine.

Shannon (1981/382grt) and *Kerne* (1913/63grt).

Manchester Ship Canal

MSC *Victory* (1974/137grt) at Irlam Loch in the Manchester Ship Canal.

MSC *Volant* (1976/137grt) also at Irlam.

MSC *Viceroy* (1975/137grt) assisting a 250-ton crane on the Manchester Ship Canal on 16 May 1992.

MSC *Talisman* (1961/124grt) in the Manchester Ship Canal on 1 May 1967. It became *Salvor II* in 1981, *Portolos Salvage* in 1988 and *Portolos Deka* in 1994 (later listed as IMO 7943110).

Daniel Adamson (1903/175grt) and MSC *Viceroy* (1975/130grt). *Daniel Adamson* was towed from Eastham to Liverpool by the *Ashgarth*. It was built by the Tranmere Bay Development Company at Birkenhead as *Ralph Brocklebank*, a tug/tender for the Shropshire Union Canal and Railway Company. It was used for many years towing barges between Ellesmere Port and Liverpool Docks. In 1921, it was owned by the Manchester Ship Canal Company, then modified eight years later and rebuilt in 1936, becoming the *Daniel Adamson*. The tug was retired in 1984 and berthed at the Ellesmere Port Boat Museum. A preservation society was formed in 2004 and work was completed on the hull, accommodation and engines. The vessel now operates cruises on the Mersey and Manchester Ship Canal.

MSC *Dainty* (1959/37grt) was built by Isaac Pimblott & Sons at Northwich and operated on towing duties for the Manchester Ship Canal Company. It is pictured here on 16 May 1992.

Northwich

Proceed (1955/126grt) at Pimblott's Old Yard, Northwich, on 15 May 1997.

Preston

Hannah Spearing ,1943, 176grt, 26.4×7.4, 700, IMO 5080225.
b. American Shipbuilding Company, Buffalo, Yard No. 922.
It was built as *ST-767*, becoming *Cornhill* in 1951 and *Hannah Spearing* in 1980. It was broken up at Fleetwood in 1999 following a serious fire in 1996.

Heysham

Gary Grey (1954/37grt) ay Heysham in September 1988.

Sea Trojan, 1962, 107grt, IMO 5316466.
b. John Lewis & Sons Limited, Aberdeen, Yard No. 320.

Douglas, Isle of Man

Salisbury (1943/54grt) and *Primrose* (1906/52grt) at the Inner Harbour at Douglas, Isle of Man. *Primrose* was built in 1906 at Appledore, Devon, and delivered to the Sharpness New Docks & Gloucester & Birmingham Navigation Company. They were both operated by the Laxey Towing Company Limited.

Workington

Beaver Crest, 1956, 64grt, 21.0×5.0.
Westminster Dredging Company Limited.
Built as *Pegasus*, it became *Beaver Crest* in 1957. It was reported that in 1984 at Humberside while towing a crane and barges it became overrun by the barges and sank. It was later raised and, following a survey, was declared a constructive total loss and scrapped.

Sovereign, 1957, 147grt, 29.2×7.9, IMO 8227226.
b. P. K. Harris Limited, Appledore, Yard No. 105.
Riverway Developments Limited.
It was built as MSC *Sovereign*, becoming *Sovereign* in 1984 and *MB 1* in 1997.

Solway, 1943, 232grt.
b. A Hall, Aberdeen.
Workington Harbour Dock Company.
Built as *Empire Anne*, it was renamed *Solway* in 1951 and broken up at Fleetwood in 1977.

Belfast

Craigdarragh, 1966, 169grt, IMO 6519302.
b. Appledore Shipbuilders Limited, Appledore. Yard No. as11.
Cory Ship Towage, Northern Ireland.
It was renamed *Ikar* in 1994.

Masterman, 2009, 247grt, 24.4×9.2, IMO 9569023.
b. Pirlant, Tuzla, Yard No. 21.
United Towing.

Svitzer Trident, 2016, 293grt, 28.7×10.4, IMO 9769398.
b. Damen, Galati, Yard No. 2630.
Svitzer Marine Limited.

Ardrossan

Seaway, 1942, 260grt, 1000 (1), IMO 5317006.
b. Scott's of Bowling Yard No. 362.
Ardrossan Harbour Company.
It was built as *Empire Palm*, renamed *Seaway* in 1946 and was broken up at Dalmuir in 1969.

Clyde

Flying Fulmar, 1974, 298grt, 38.2×9.4, 2400 (37). IMO 7324118.
b. Ryton Marine, Wallsend, Yard No. 529.
Clyde Shipping Company Limited.
It was renamed *Alce Nero* in 1992.

Anglian Knight, 1975, 665grt, IMO 7403483.
b. Jonker & Stans, Hendrik-Ido-Ambacht, Yard No. 332.
Awal Contracting & Trading.
Built as *Union One*, it became *Anglian Knight* in 1988, *Charger 2* in 1989, *Awalco 4* and *Challenger* in 1990. It was broken up in 2019.

Roysterer, 1972, 1036grt.
b. Charles D. Holmes, Beverley.
Royal Fleet Auxiliary (RFA).
The Roysterer class were 1,630 tons displacement, 179.7 feet overall, diesel engines, twin screw, 4,500 bhp, 15 knots, oil fuel 93 tons, range 12,300 nautical miles, crew thirty-one, bollard pull 50 tons.

Thunderer, 1970, 272grt, IMO 6929997.
b. J. Lamont & Company Limited, Port Glasgow, Yard No. 411.
Cory Ship Towage (Clyde).
It was renamed *Thunderer* in 1973 and *Iron Bull* in 2003.

Caledonian Canal

Martello, 1958, 68grt, 440(B) M.
Leith Dock Commissioners.
It is seen in the Caledonian Canal in October 1996.

Aberdeen

Sea Trojan (1962/117grt) and *Sea Griffon* (1962/117grt) operated for the Aberdeen Harbour Board.

Grangemouth

Kelty, 1976, 322grt, 38.0×9.6, IMO 7392684.
b. Richards (Shipbuilders) Limited, Lowestoft, Yard No. 526.
Forth Tugs Limited.
It was renamed *Shireen S* in 1998 and *W Power* in 2013.

Almond (1976/322grt) and *Duchray* (1975/326grt) operated for Forth Tugs Limited at Grangemouth.

Boquhan, 1975, 326grt, 38.2×9.7, 2640 (37), IMO 7392828.
b. Robb Caledon Limited, Leith, Yard No. 518.
Forth Tugs Limited, Hound Point.
It was renamed *Oliver* in 1997 and *Nano* in 2009. It was broken up in 2009.

Blyth

Mister Cornishman, 1959, 230grt, 1680 (B), IMO 5144899.
b. P. K. Harris Limited, Appledore, Yard No. 111.
Western Ocean Towage Company Limited, Plymouth.
It was built as *Hazelgarth*, becoming *Mister Cornishman* in 1989, *Shaula* in 1990 and *Tommy Dev* in 1994.

River Tyne

Cullercoats (1898/181grt) and *Ashbrooke* (1955/119grt).

Flying Spindrift (1986/259grt), *Northsider* (1967/156grt) and *Hillsider* (1988/156grt).

Tynesider (1942/262grt).

Tynesider (1942/262grt), *Joffre* (1916/260grt) and *Beamish* (1944/242grt) working on the River Tyne on 15 June 1965.

Anglian Duke (1977/762grt) with *Tuxedo Princess* in the River Tyne on 19 May 1998.

Torque, 1943, 232, IMO 7606346.
b. A Hall & Company Limited, Aberdeen, Yard No. 690.
Thames Services (Marine) Limited.
Built as *Empire Spitfire*, it became *Warden* in 1947, *Prompt* in 1951 and *Torque* in 1975.

Sunderland

Fullwell (1899/149grt) and *Houghton* (1904/133grt) were operated by France, Fenwick Tyne & Wear Company Limited.

Dunheron, 1955, 133grt, IMO 5133034.
b. Scott's of Bowling, Yard No. 407.
It was built as *Golden Cross*, becoming *Dunheron* in 1968 and *Golden Cross* again in 1997. It was broken up in 2013.

Wearsider, 1964, 151grt, 986 (B), IMO 7818000.
b. Richard Dunston Limited, Hessle, Yard No. h929.
Wear Tugs Limited.
It became *Samurai Swordsman* in 1991, *Wearsider* in 1992, *Megalochari VI* in 1995 and *Artemis V!* in 2017.

Holmsider, 1984, 164grt, 28.0×7.8, IMO 8320597.
b. Richard Dunston Limited, Hessle, Yard No. h943.
Wear Tugs Limited.
It became *Wyegarth* in 1997, *Karapiperis 15* in 1998 and *Evripos IV* in 2019.

Seaham

Seaham (1909/133grt) and *Reliant* (1907/156grt).
Seaham Harbour Dock Company.

Hartlepool

Formidable, 1979, 422grt, 35.7×10.3, IMO 7719698.
b. Richard Dunston Limited, Hessle, Yard No. Y917.
Alexandra Towing Company.
It became *Eurosund* in 2001, *Mignon* in 2010, *Hamal* in 2013, *Formidable* in 2016 and *BVT Formidable* in 2020.

MCS *Rapier*, 1983, 132grt, IMO 8207977.
b. Richard Dunston Limited, Hessle, Yard No. 936.
Maritime Craft Services (Clyde) Limited.
It became MCS *Rapier* in 1994 and *Lachay* in 1995.

Stranton (1959/145grt) and *Hart* (1958/145grt).
Both tugs were built by P. K. Harris Limited, Appledore, Yard Nos 124 and 112.
Delivered to British Transport Docks, which later became Tees & Hartlepool Port Authority.

Towing Wizard, 1955, 137grt, 29.6×7.1, 1200 (16), IMO 5287823.
b. Mutzelfeldewerft Nord, Cuxhaven, Yard No. 139.
TSA Tugs Limited.
It was built as *Quaysider*, becoming *Tora* in 1994 and *Nafpigia Panagiotaki* in 2004.
Towing Wizard is shown assisting *Wingfield Castle* (1934/556grt) on 20 May 1986.

Tees

Jerome Letzer (1974/304grt) and *Tuxedo Royale* enter Middlesbrough Dock on 31 March 2000.

Smith's Dock No 2, 1939, 30grt.
b. Smith's Dock Company Limited, Middlesbrough, Yard No. 1057.
Shown here on 2 September 1975. It was broken up in 1981.

Tees tugs awaiting orders on 18 February 1985.

Ralph Cross (1974/244grt) becomes *Crez* following its sale in 1989.

Coatham Cross, 1981, 185grt, 27.8×8.9, 2640 (35), IMO 7928043.
b. Richard Dunston Limited, Hessle, Yard No. 931.
Tees Towing Company Limited.
It was renamed *Cormilan* in 2011 and *La Dani* in 2014.

Vanguard (1965/306grt) towing gear at Middlesbrough on 16 November 1994.

Ayton Cross's (1967/214grt) towing hook just prior to demolition at Hartlepool in 1985.

York

France Hayhurst at York on 23 April 1991.

Tilbury, also at York, on 7 March 1992.

Hull

Foreman, 1959, 227grt, 32.6×8.08, 1200 (1) SR, IMO 5117585.
b. Cook, Welton & Gemmell Limited, Beverley, Yard No. 935.
United Towing Company, Hull.
Wrecked at Pontevedra Bay on 2 October 1981.

Lady Theresa, 1988, 156grt, 24.3×7.7, IMO 8616245.
b. Cochrane Shipbuilding Company, Yard No. 136.
Humber Tugs Limited, Immingham.
It became *Hillsider* in 1994, *Fairplay XI* in 2001 and *Twister* in 2013.

Norman, 1929, 222grt, 1100 (1) M, IMO 5256214.
b. Cochrane & Sons Limited, Selby, Yard No. 1047.
United Towing Company Limited.
Between 1939 and 1945 it became *Diversion*, then *Norman* again in 1945. It was converted into a motor ship in 1967 and was broken up at Hull in 1972.

Presto, 1943, 276grt, 1000 (1) SR.
b. Cochrane & Sons Limited, Selby, Yard No. 1267.
Ellermans Wilson Lines Limited.
Built as *Empire Sara* and became *Presto* in 1946. It was broken up at Blyth in 1968.

Redbridge (1995/399grt) fitting out at Hull.

Lady Joan (1987/156grt) soon after launch at Cochrane Shipyard, Selby. It became *Conero* in 1992.

Allerton Bywater

British Waterway tugs at Allerton Bywater, West Yorkshire.

Immingham

Lady Josephine, 1991, 364grt, 30.6×11.5, 4800, IMO 8919219.
b. McTay Marine Limited, Bromborough, Yard No. 94.
Humber Tugs Limited.
It became *Adsteam Josephine* and *Svitzer Josephine* in 2007.

Great Yarmouth

Holyhead Towing Company's *Afon Wen* (1988/44grt) at Yarmouth on 15 January 1995.

Left: Funnel markings on *Hector Read* (1965/67grt) in September 1986.

Right: *Hector Read*, 1965, 67grt, 19.8×5.8, 545.
Great Yarmouth Port & Haven Commissioners.
This was the last vessel to be built at the shipyard of J. S. Doig in Grimsby. It worked at Great Yarmouth for thirty-five years before being purchased by Whiting Shipping.

Lowestoft

Eta, 1955, 52grt, 21.5×5.6, 360.
Colne Shipping Company Limited, Lowestoft.

Ala, 1961, 63grt, 20.7×6.3, 400.
Colne Shipping Company Limited, Lowestoft.

Ipswich

Maximus, 1956, 141grt, 29.6×7.1, 570, IMO 5230038.
b. Mutzelfeldewerft Nord, Cuxhaven, Yard No. 144.
M. Tugs Limited.
Broken up at Lowestoft in 2011.

Felixstowe

Grey Vixen, 1991, 40grt, 16.5×5.2, 850, IMO 4501002.
b. IHC Deltawerf, Sliedrecht, Holland.
Felixarc Marine Limited, Felixstowe.

Harwich

Brightwell (1986/256grt) and *Ganges* (1982/281grt) at Harwich.

Maldon

Built by William Pickersgill & Sons of Southwick, Sunderland, *Brent* (1945/54grt) is the last surviving steam vessel of the Port of London Authority's fleet and one of three vessels left of those used on the River Thames. It is a classic small lighterage steam tug and was saved from scrapping in 1970 by the Hall family of Maldon.